3724

791.45

D1143345

QUEENS GATE SCHOOL
133 QUEENS GATE
LONDON SW7 5LE
TEL 071-589 3587

WITHDRAWN

...EENS GATE SCHOOL
33 QUEENS GATE
LONDON SW7 5LE
TEL 071-589 3587

WITHDRAWN

QUEENS GATE SCHOOL
133 QUEENS GATE
LONDON SW7 5LE
TEL 071-589 3587

QUEENS GATE SCHOOL
133 QUEENS GATE
LONDON SW7 5LE
TEL 071-589 3587

MEDIA FOCUS

Television

David Self

QUEENS GATE SCHOOL
133 QUEENS GATE
LONDON SW7 5LE
TEL 071-589 3587

QUEENS GATE SCHOOL
133 QUEENS GATE
LONDON SW7 5LE
TEL 071-589 3587

NS GATE SCHOOL
3 QUEENS GATE
ONDON SW7 5LE
TEL 071-589 3587

Heinemann
LIBRARY

First published in Great Britain by Heinemann Library
Halley Court, Jordan Hill, Oxford OX2 8EJ
a division of Reed Educational and Professional Publishing Ltd.
Heinemann is a registered trademark of Reed Educational & Professional Publishing Ltd.

OXFORD MELBOURNE AUCKLAND
KAMPALA JOHANNESBURG BLANTYRE GABORONE
IBADAN PORTSMOUTH (NH) USA CHICAGO

© Reed Educational and Professional Publishing Ltd 1998
The moral right of the proprietor has been asserted.

All rights reserved. No part of this publication may be reproduced, stored in a retrieval system, or transmitted in any form or by any means, electronic, mechanical, photocopying, recording, or otherwise without either the prior written permission of the Publishers or a licence permitting restricted copying in the United Kingdom issued by the Copyright Licensing Agency Ltd, 90 Tottenham Court Road, London WIP 0LP.

Designed by Jim Evoy
Illustrations by Jeff Edwards
Printed in Hong Kong

02 01 00 99
10 9 8 7 6 5 4 3 2

ISBN 0 431 08247 2

British Library Cataloguing in Publication Data
Self, David
      Television, - (Media focus)
      1.Television - Juvenile Literature
      Title
      302.2'345

Acknowledgements
The Publishers would like to thank the following for permission to reproduce photographs:
Anglia TV pp.6, 13, 16 (bottom right), 19 (top); Bazar TV, p.22; BBC pp.4, 20 (right), BBC/Aquarius p.22 (left); BSkyB p.29 (bottom right and logo), Sam Teare p.16 (bottom); Carlton Television Ltd, p.14; Central TV, p.7; Channel 4, p.5; Channel 5, p.5; ITC p.25; ITN, p.9; Jeff Moore p.29 (top right); Lambie-Nairn p.27; L!ve TV, p.8; Pictorial Press Ltd/T McGough p.20 (centre); Rex Features, (Chris Brown), p.11, (Jonathan Buckmaster), p.24; Rooke & Jorgensen p.11; Sporting Pictures p.16 (top right); UK TV, p.29.

Cover photograph reproduced with permission of Astra.

Our thanks to Steve Beckingham, Head of Media Studies, Fakenham College, Norfolk, and to Paul McGhee for their comments in the preparation of this book.

Every effort has been made to contact copyright holders of any material reproduced in this book. Any omissions will be rectified in subsequent printings if notice is given to the Publisher.

Any words appearing in the text in bold, **like this**, are explained in the Glossary.

Contents

Introduction

Despite the fact that people watch less television than they did ten years ago, it is still the favourite spare time occupation of most people. 40% of all our leisure time is spent watching television. Television is a huge industry. This book is an introduction to the way that industry works in Britain and the ways factual television programmes get made.

Who owns the channels?

Always watch ITV? Never watch the other side? Always zapping channels? Who owns these different stations – and who pays for them?

BBC

The BBC is seen by many people in this country and abroad as the world's leading broadcasting organization. Over the last 75 years it has gained a reputation for high-quality broadcasting on both television and radio. When a major news **story** breaks or during other important events (such as a royal funeral), it is the channel most people watch. 95% of all households tune in to the BBC for at least two hours a week.

The BBC is a public service broadcaster: it provides a service for the public without showing adverts and without having to make a profit. This role is made possible by the television licence fee. Every household with a television set must pay the licence fee, whether they watch BBC programmes or not. In December 1996, the government announced a settlement that fixed the colour television licence fee at £97.50 per household per year until the year 2002.

The boss of the BBC is its Director-General. But just as a school headteacher is responsible to a board of governors, so is the BBC's Director-General. The BBC Governors in turn have a duty to ensure that the licence payers receive the best value from the licence fee, and to see that BBC programmes inform, educate and entertain.

BBC ONE

BBC 1 is meant to appeal to as many people as possible. Every year, it broadcasts features, documentaries and current affairs programmes, light entertainment or comedy programmes, 'soaps', films and other dramas, sport and children's programmes. It also carries the main **news programmes** and the weather forecasts.

BBC TWO

BBC 2 carries programmes that will appeal to minorities within the general audience. Its comedy programmes often appeal to younger people than those on BBC 1. It also broadcasts serious music and arts programmes, challenging drama, adult education, community, **special needs** and schools programmes.

Programmes seen on BBC Television are made by London-based production departments and also by the six BBC regions based in Scotland, Wales, Northern Ireland, Birmingham, Bristol and Manchester. BBC Television must also buy 25% of its programmes from independent producers. Feature films, mini drama series, cartoons and overseas sport are bought by BBC Television from other organizations around the world. The BBC also sells programmes to other countries and to **satellite** and **cable** channels. Profits are used to make more programmes.

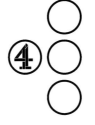

1	Carlton/LWT	8	Anglia
2	Meridian	9	Harlech
3	Central	10	Westcountry
4	Granada	11	Channel
5	Border	12	Scottish
6	Yorkshire	13	Grampian
7	North East 3 (Tyne Tees)	14	Ulster

Areas where 2 or 3 stations can be seen

ITV regions and their overlap areas.

Independent Television (also known as ITV or Channel 3) consists of fifteen regional companies and one other separate company, GMTV, which provides a nationwide breakfast programme. In the London area, ITV is provided jointly by Carlton Television (weekdays) and London Weekend Television. These sixteen companies won licences from the Independent Television Commission (ITC) to provide regional services for a ten-year period from 1993. After that, they will have to compete for new licences.

ITV companies are owned by their shareholders. While they have a duty to

the ITC to make certain types of programme, they also have to make money for their shareholders. They make profits by selling programmes to each other and to other broadcasters, but especially by selling screen-time to advertisers.

In recent years, some of the bigger ITV companies have been buying up the smaller ones. At the end of 1997, Carlton owned Carlton, Central and Westcountry Television and part of GMTV. Granada owned Granada, LWT, part of Yorkshire Television and also a share of BSkyB (British Sky Broadcasting). A company called United News & Media owned Anglia, Harlech and Meridian Television and a share of Channel 5. For satellite and cable providers, see pages 28–29.

see pages 28–29.

DID YOU KNOW? FACTS & FIGURES INDUSTRY STATISTICS VITAL STATISTICS DID YOU KNOW? FACTS & FIGURES INDUSTRY STATISTICS

- CHANNEL 4 BECAME A FULLY INDEPENDENT PUBLIC BROADCASTING SERVICE IN 1993.
- LIKE ITV, IT EARNS MONEY BY SELLING ADVERTISING TIME, BUT ITS PROFITS ALL GO TOWARDS MAKING NEW PROGRAMMES, NOT TO SHAREHOLDERS. IT HAS A DUTY TO MAKE TYPES OF PROGRAMME NOT USUALLY SEEN ON ITV, AND 'TO ENCOURAGE INNOVATION AND EXPERIMENT'. BECAUSE OF THIS, SEVERAL OF ITS PROGRAMMES HAVE SHOCKED SOME PEOPLE.
- CHANNEL 4 BUYS ITS PROGRAMMES FROM INDEPENDENT PRODUCERS.
- IN WALES, THE FOURTH CHANNEL CARRIES THE PROGRAMMES OF S4C (SIANEL PEDWAR CYMRU), WHICH IS A MIXTURE OF WELSH LANGUAGE PROGRAMMES (SOME MADE BY THE BBC) AND CHANNEL 4 PROGRAMMES.

- CHANNEL 5 IS AN INDEPENDENT, COMMERCIAL COMPANY.
- LIKE ITV AND CHANNEL 4, IT EARNS MONEY BY SELLING SCREEN-TIME TO ADVERTISERS.
- LIKE THE ITV COMPANIES, IT GIVES ITS PROFITS BACK TO ITS OWNERS OR SHAREHOLDERS.
- LIKE ITV, IT IS CONTROLLED BY THE ITC.

The television studio

Sound on! Vision on! Then you press the wrong button and eight million viewers laugh at your mistake. That's live television.

The gallery

The **gallery**, or control room, is often higher than the studio floor. It may be possible to see into the studio, but everyone in the gallery relies on the **monitor** screens in front of the control desk to see what's going on.

If there are five cameras involved in the production, then five of these screens will show the output of those cameras. Another will show any captions that are being created electronically, another any slides, and yet another shows any film or pre-recorded video inserts that are played into the

production. The central screen shows which picture or combination of pictures is being transmitted or recorded.

At the centre of the desk sits the **director**. Besides watching the monitors, he or she can hear what is said in the studio, and can also speak to and hear other production departments, the **floor manager** and the technicians.

On one side of the director is the **vision mixer**, who sits in front of a complex control

panel. The vision mixer has to watch all six or seven monitor screens and cut, fade, wipe or mix from one source to another at exactly the moment that was decided in advance and was noted in the camera script, which he or she is now too busy to read.

On the other side of the director is the **production assistant**, who is responsible for time-keeping and for making sure the production ends on time. He or she notes all the 'takes', changes to the script and timings for use during **editing**.

Next to main control is sound control. In charge is the sound supervisor, who opens and closes each microphone on the studio floor. He or she is responsible for the technical quality of the sound.

On the other side of the main control is lighting control. Here the lighting engineers operate a console that controls all the studio lights. Also in lighting control – or racks, as it is traditionally known – are the engineers who are responsible for adjusting and controlling picture quality from the cameras.

The studio floor

In the centre of the studio is the set for the programme being broadcast or recorded. The programme designer will have worked closely with the director to create a set that is practical and which suits the mood of the programme.

There might be four, five or six cameras. Most can glide smoothly around the studio. Occasionally a camera is mounted on what is basically a small crane to give it greater movement and height. On top of each camera is a tiny monitor screen, in which the camera operator can see the shot he or she is offering to the director and vision mixer. A red light on top of each camera shines when that camera's shot is being used. There is often a boom microphone. This microphone is suspended from a long extendable pole, or boom, itself mounted on a mobile platform (sometimes called the boom pram). It can be moved in various ways to catch the comments from, say, a studio audience. Other speakers use fixed microphones built into the set, or radio microphones pinned on their lapel or neckline. A tiny hidden transmitter sends the sound signal to sound control. Hanging from the ceiling is a close-packed array of lamps. Access to this lighting-grid is possible via catwalks.

The floor manager

The person in charge of the studio floor is the floor manager. He or she wears a two-way headset which allows them to hear what is being said in the gallery and (when they depress a switch) to speak to the director. The floor manager must ensure everything in the studio runs smoothly, calmly and on time. A good floor manager is strict, good-humoured and tactful. For example, when the director yells in the floor manager's **earpiece**, 'For the tenth time, tell that little creep not to sit on the edge of the sofa like a constipated chimp', the floor manager politely suggests that the speaker in question should lean back in the sofa 'just for the cameras'.

Besides the camera operators (and their assistants, who may be required to move cables at particular moments), there are electricians, set dressers (responsible for adjusting the set) and make-up assistants hovering around the studio.

DID YOU KNOW?•DID YOU KNOW?•DID YOU KNO

Sound control

TYPICALLY, A SOUND SUPERVISOR IS RESPONSIBLE FOR:

- TWO OPEN REEL AUDIO TAPE RECORDERS
- A DIGITAL TAPE RECORDER (DAT)
- A CASSETTE TAPE RECORDER
- A CD PLAYER
- UP TO EIGHT RADIO MICROPHONES
- FIXED MICROPHONES
- A BOOM MICROPHONE
- SEVERAL RADIO TALKBACK CHANNELS (THE AUDIO LINKS BETWEEN GALLERY AND STUDIO FLOOR)

The news agenda

'It's a great story. 250 dead. Should be the lead item.' 'Yes, but it's in a two-bit banana republic and we haven't got any pictures.'

The editors of television **news programmes** want to report what is going on in the world accurately, fairly and quickly. After all, over 71% of people regard television as their first source of news, and they expect to be able to trust what they see and hear.

On the other hand, news editors don't want viewers to get bored and switch off. If they do, network controllers might start thinking that less money should be spent on the news, or that news programmes should be moved to less popular times. It has been calculated that by replacing ITV's *News at Ten* by more popular programmes such as films and 'adult dramas', ITV might earn £50 million in extra advertising.

So editors often class news stories under one of these headings:

1 Important and interesting
2 Important but not interesting
3 Interesting but not important
4 Neither interesting nor important

L!ve TV, a London cable station, wants to be sure the news is always interesting. Its news bunny gives each story the thumbs-up or thumbs-down.

News values

What goes in the news is not just a matter of pleasing the viewer. These are some of the things that editors and journalists think make an event newsworthy:

• Importance. If an event is thought to be trivial, it will not be reported. When a family gets a new pet dog, that isn't 'news'. If the new pet was a gorilla, then that might be news. But small incidents can lead to bigger events. For example, a small cut in what a town spends on street lighting could lead to more muggings and traffic accidents. News teams have to remember that little events might one day become important.

• Simplicity. Some events – such as political arguments or financial news – may be so complicated that they are very difficult to report in 90 seconds. This can mean they don't make it into some **bulletins**.

• Topicality. What happened three days ago, or what might happen in a year's time, is not news.

• Relevance. A story is news if it matters to the average viewer. Often a story is reported in terms of how it affects the public. For example, reports of the Government's annual spending plans (the Budget) often concentrate on what they mean to a typical family – such as rises in the price of beer and petrol.

• Continuity. Once an event has become a story, new developments are likely to be reported.

For example, when protesters are trying to stop the building of a new motorway, each day's events become an episode in the on-going story.

• Balance. Most news organizations try to achieve a good balance of news. For example, they will try to provide equal amounts of news about the different political parties – especially at election time – and they will also try to balance good and bad news. Editors don't usually like a bulletin to contain only gloomy stories.

• Personalization. Stories are often personalized. 'The Minister for Health has decided that three London hospitals must close' is more newsworthy than 'The Government has decided that savings of £3 million can be made by re-organizing London's healthcare'.

• Rich and famous. Some people seem to be so rich and famous that whatever they do gets reported.

• Negativity. Crashes, deaths, natural disasters (earthquakes, floods, etc.) and other negative happenings always attract attention.

Intrusion

Should the news show private moments?

In 1984, the IRA bombed a Brighton hotel where many leading politicians were staying. An injured cabinet minister, Norman Tebbit, was televised being rescued from the rubble on a stretcher in his pyjamas. Later, he said that he had no objections to the broadcast because it made people think how terrorists caused human suffering.

But news editors are sometimes accused of going too far. In 1989, terrorists exploded an aircraft over Lockerbie in Scotland. Some television news programmes showed the mother of one victim falling on the floor at the airport where she had hoped to meet her daughter, and screaming with grief.

Later she said, 'Everything in my life had been taken from me. I wasn't even allowed to experience the pain of that moment without someone intruding.'

ITN is the company which provides news programmes for the non-BBC networks. It is owned jointly by Carlton Television, Granada, the *Daily Mail*, United News and the newsgathering organization Reuters. This is how it describes some of its main daily programmes:

5.40-6.00pm Early Evening News
Subtitled **Weather** *391861*

'crisp…fast…concise'

7.00pm Channel 4 News
With Jon Snow and Cathy Smith.
Weather *Stereo Subtitled 824126*

'takes a deeper longer look'

8.30pm 5 News
Including *First on Five*.
Stereo Subtitled *824126*

'fast moving…aimed at a younger audience'

10.00pm News at Ten
With Trevor McDonald.
Weather Sian Lloyd
Subtitled *41478*

'substantial'

Which of the four types of news story (see opposite page) do you think will feature most in each of these programmes?

Newsgathering

A man in a suit sits at a desk reading the news. But where did he get the news from? Who gathers the news – and how?

Television news never stops. On some channels there may be just three or four news **bulletins** a day. Other channels are 'rolling news' channels with non-stop news programmes. In either case, the newsroom that produces those bulletins works around the clock.

The first broadcast is usually in the early morning as the breakfast news programme (prepared during the night) goes on air. At the same time, the duty assistant news editor arrives to organize the new day's coverage. He or she will already know about some of the **stories** to be covered. Political conferences, courtroom trials and protest meetings are planned in advance. Details will be in the newsroom's news diary, which lists all known future events. Reporters and camera crews will have been assigned to each story that is to be covered.

Breaking the news

During the day, news will come in of breaking stories – that is, unpredicted events that suddenly happen. First will come stories from countries to the east, where the day has started before ours. For example, a reporter (or correspondent) in Israel may alert the **intake editor** to a riot starting in Jerusalem. As the day goes on, home news (news about Britain) will come from a variety of sources: phone calls, the emergency services, regional newsrooms, the House of Commons... Later in the afternoon, as America wakes up, stories will begin to come in from New York and Washington. In every case, the intake editor must decide whether to cover the story – and how. Where is the nearest reporter? Where is the nearest camera crew? (Because they use video, these are known as **electronic newsgathering** or **ENG** crews.)

A reporter may cover a story with just one cameraperson (who also records the sound). In other cases there may be a separate sound engineer and someone to arrange additional lighting. Sometimes, the recorded videotape will be sent back to the studio by motorcycle despatch rider. At other times, a **satellite** link will be set up.

Meanwhile, researchers will be contacting experts to be interviewed, and specialist correspondents (reporters who concentrate on one area of news such as politics or the environment) will be preparing other news items.

Editor

Assistant News Editors

Intake Editor | Output Producer

Newsdesk | Forward Planning

Specialist Correspondents

Facility Engineers

Reporters

Camerapersons

'We'll do you down the line'

There are many small interview studios around the country, remote from the main studios where news programmes are presented. They are connected by telecommunications links known as lines.

When there isn't time to get a reporter and camera to someone in the news, that person may be asked to go to the nearest remote studio. A technician or researcher shows them into the studio, switches on the equipment and makes contact with the main news base. The interviewee hears the interviewer either on a tiny **earpiece** or on a speaker in the studio, but cannot see who he or she is talking to.

When someone is being interviewed face to face by a reporter, the interviewee is asked to look at the reporter. When being interviewed 'down the line', the interviewee is usually asked to talk directly to the camera. An interview down the line may be recorded for later use, or fed live into a **news programme**.

On the satellite link

Quite often, a reporter will have to give a live report by satellite directly into a news bulletin.

Imagine the possible situation. You have spent the day in court, listening to a murder trial. You are now standing in a car park outside the court house, with a camera operator and a van with a satellite dish on it. This 'links' vehicle will bounce your report up into space and then down to the main studio.

You stand facing your camera listening to the studio control room on your earpiece. 'Just stay there,' they tell you. 'We could come to you at any time.'

The local children are beginning to think it is a great idea to leap up and down behind you, pulling faces. Suddenly you find yourself on air, beamed into a million homes, reporting on the day's events. You must report what happened in court precisely and in exactly 50 seconds. Sometimes a dozen news teams may report the same event.

DID YOU KNOW?•FACTS & FIGURES•INDUSTRY STATI

- BBC TELEVISION NEWS HAS ACCESS TO 100 ENG CREWS.
- BBC TELEVISION NEWS HAS 453 REPORTERS AND 66 SPECIALIST CORRESPONDENTS IN BRITAIN. IT HAS 22 FOREIGN CORRESPONDENTS AROUND THE WORLD.
- ITN HAS 70 HOME REPORTERS WITH OFFICES (OR 'BUREAUX') AND CORRESPONDENTS IN WASHINGTON, MOSCOW, BRUSSELS, JOHANNESBURG, JERUSALEM, HONG KONG AND BEIJING. IT ALSO HAS REGIONAL BUREAUX IN BRITAIN.
- SKY NEWS PROVIDES A 24-HOUR NEWS SERVICE FROM LONDON FOR MUCH OF EUROPE. ITS HOURLY BULLETINS ARE SEEN BY 70 MILLION VIEWERS IN 40 COUNTRIES.
- THE AMERICAN CNN (CABLE NEWS NETWORK), WHICH IS SEEN WORLDWIDE, HAS BUREAUX (WITH REPORTERS) ALL AROUND THE WORLD. ITS LONDON PRODUCTION CENTRE ORIGINATES FOUR HOURS OF BROADCASTING EVERY 24 HOURS.

Here is the news

'Why don't you smile more often?' That, say many news presenters,
is the question they are most often asked about their work.

In America, they are known as anchors. In Britain, they are called newsreaders, newscasters or **presenters**. Once, they were chosen for their beautiful voices – and because they looked good on screen. Now, most of them are journalists. But as competition grows between the different channels, some people claim that looks are again becoming important. No one wants to watch ugly newsreaders – no matter how good they may be as journalists.

Newsreaders are expected to appear confident and to have authority, but not to seem cocky; to be friendly but not too cosy; to be impartial but not too cold; and never to let their own views colour the news. Some, however, may see themselves as personalities. In either case, they are still just one part of the team that gets the news on screen.

Before the invention of videotape, television news reporters were photographed on film. Once a reel of film had been shot on location and brought back to the studio, it had to be processed (like the film in an ordinary camera) before it could be **edited** – which involved cutting the actual film. Video is much quicker. Pictures can be edited as soon as the tape arrives in the newsroom. It is not cut. Instead, the required parts are dubbed, or copied, and then re-assembled, while any necessary **voice-over** is added to help tell the **story**.

Modern newsrooms are digital. The raw video material from the reporters and other sources is fed into the newsroom's computer system and is immediately available to the editors of all the various **bulletins** and news programmes, and to the editor of the rolling news service, if there is one (see opposite). Computer or digital editing is even quicker than dubbing tape.

The editor of a news programme (or, often, the chief sub-editor) decides the running order of the programme and works out the timing of each item. The other sub-editors are responsible for re-writing **links** provided by the reporters, and for writing the rest of the newsreader's script.

Working alongside the editorial team are several other departments. For example, most newsrooms have an extensive VT (videotape) library which allows news teams to use pictures from old news reports to help explain and illustrate current stories.

Editor

Intake Editor — Output Producer

Studio Director — Programme Editor

Presenters — Studio staff (sound, cameras lighting, etc.) — Sub-Editors

Journalists

News Assistants — Graphics

VT and Stills Archivists

Job titles vary from one news organization to another.

The BBC has a Graphic Design Department with a hundred staff. They provide graphics (captions, charts, diagrams, etc.) and still pictures. Most graphics are created on electronic paintboxes, using an electronic stylus and tablet. Others are created digitally by a computer. The images are formed directly as video, not as artwork on paper. Three-dimensional images can also be created in these ways.

The BBC has over a million stills, consisting mainly of 35mm colour slides or black-and-white photographs. These include a huge collection of world location and portrait shots covering news, current affairs and sport. As these are now being kept on optical disc, it can take less than five seconds to call up a picture of a place or person who suddenly comes into the news.

As here, the presenter's desk is often long enough to seat two newsreaders or even two readers and a reporter.

In the studio

A television news studio is much like other television studios. In the **gallery** there will be a senior engineer or technical co-ordinator, the programme **director**, **production assistant**, **vision mixer** and sound operators. In all news operations there is a move towards multi-skilling, where one person is expected to do several jobs.

Because a news presenter does not normally move about, news studio cameras do not have to move from one position to another. They can therefore be operated remotely (or robotically) from the gallery by just one person. Another key person on the studio floor is the **autocue** operator.

DID YOU KNOW?•FACTS & FIGURES•INDUSTRY ST

Rolling news

A TELEVISION CHANNEL WHICH CARRIES NOTHING BUT NEWS IS CALLED A ROLLING NEWS SERVICE. THE MOST FAMOUS IS CNN. THE FIRST BRITISH ROLLING NEWS 24-HOUR SERVICE WAS SKY NEWS, WHICH IS SEEN BY OVER 70 MILLION VIEWERS IN OVER 40 COUNTRIES. THE BBC ALSO PROVIDES A WORLDWIDE 24-HOUR NEWS AND CURRENT AFFAIRS SERVICE CALLED BBC WORLD, WHICH IS NOT SHOWN IN BRITAIN.

SINCE 1997, THE BBC HAS ALSO BROADCAST A ROLLING NEWS SERVICE FOR UK VIEWERS: BBC NEWS 24. IT DRAWS HEAVILY ON THE BBC'S THIRTEEN REGIONAL NEWSROOMS, SO IT CONTAINS MUCH MORE NEWS ABOUT BRITAIN THAN SKY NEWS DOES.

AT FIRST, BBC NEWS 24 COULD BE SEEN ONLY BY **CABLE** TELEVISION VIEWERS, ALTHOUGH IT HAS ALSO BEEN CARRIED ON BBC 1 AT NIGHT. IT WILL GAIN MORE VIEWERS AS MORE PEOPLE BEGIN TO WATCH DIGITAL TV.

In your region

'The trouble with the local news is there's never anything about our town on it. It's not local enough.'

This is the most common complaint about regional **news programmes** – but then they are regional, not local. Apart from one or two experiments, there are no really local television programmes in Britain. Both the BBC and ITV regions cover large areas, containing many towns. It can be difficult for news crews based in the production centre to cover stories from the other side of the region.

Some ITV companies have tried to overcome this by having separate editions of their evening news programmes for different parts of their region. As early as 1981, Television South (TVS), which then held the licence for southern England, began broadcasting separate editions for the east and west of its region. Anglia, Central, Meridian and Yorkshire are just some of the companies which transmit sub-regional news programmes. These are sometimes known as local opts or opt-outs.

Central News

The ITV Central region covers a quarter of England and, in area, is the largest ITV region. Central provides separate news programmes from Birmingham, Nottingham and Abingdon for the west, east and south of the region.

When it first went on air in 1982, there was only one service for the whole of the region, with just a handful of crews. Now 20 crews are out each day shooting tape, backed up by **satellite** links and a production staff of 200.

In each of the three newsrooms, journalists can call up video pictures on their desktop computers and **edit** them, even if someone else is using them on another terminal at the same time. They can then be fed in digital form through to the graphics design department for further editing before transmission.

Integrated news

Some people believe it is silly to have separate national and regional news programmes. They believe they should be integrated so that your regional programme includes foreign, national and local news. They say this would avoid repeating some

Sometimes viewers see shots of the newsreaders on either a still picture or tape of the story they are reporting. Some stations show the newsroom itself behind the presenters.

stories in both programmes. It would allow regional news editors to concentrate on those stories they think would interest their viewers most and allow them to include a local angle on a national story. For example, when the British Army was fighting in Bosnia, it would have allowed a regional news programme to focus on what a local regiment was doing there.

Other people say this would all be a waste of time and money, duplicating what the London newsroom could do very much better.

As Scotland gets used to having its own parliament, it is likely that Scottish viewers will want more news programmes to be made in Edinburgh and Glasgow.

Channel Island News

The Channel Islands form the smallest ITV region. There are no separate BBC Television programmes for the islands – except for a short **bulletin** after the main evening news on BBC 1. This is written and read by a BBC Radio Jersey reporter.

After you have finished your day's radio work, you drive five miles to the tiny television studio, which is inside an ordinary domestic garage. You switch on the mains, turn on the lights and fit the script into the **autocue**. You then sit at the desk, facing the camera. You are in the correct position when your nose coincides with a felt-tip mark on a nearby monitor screen.

You watch the national news. When it ends, you play a sound tape to introduce yourself and then switch yourself into vision. You have to work your own autocue with a foot pedal, and keep an eye on the clock. Your bulletin must end at the precise second that thirteen other regional bulletins are ending in other parts of the country, ready for the weather forecast.

You then turn the studio off and drive back to BBC Radio Jersey to read the 10.00 pm radio bulletin. It's less fun on foggy or snowy nights.

This map shows the BBC regional centres responsible for broadcasting news in their area. Those marked * also produce programmes for the networks

Sport

Sport is good for you. You sit there, relaxing on the sofa, occasionally cheering at the screen. Coach potato country.

There is more effort than this involved in getting a sporting event onto the screen. To cover a football match there will be a minimum of five cameras – ten for a major match. Sky Sport is famous for its elaborate coverage and has been known to use 23 cameras to cover one match.

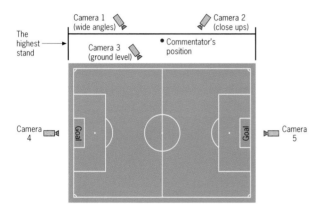

BBC commentator, Hamilton Bland, with lip mic.

you can rehearse camera movements for a horse race because you know where the horses will be going, you can only hope to follow the action at a football match.

As well as the scanner, there will be a **VTR** (video tape recording) truck with at least three video engineers. There will also be a caption generator in this van. This creates, or generates, captions such as score lines and players' names that are shown on the screen. Another van will have been used to bring the cameras and cables to the ground. Most football grounds have permanent cable or radio links to the nearest television station. In other cases, a **satellite** dish is used to send back the signal.

The **outside broadcast (OB)** crew consists of one camera operator per camera, plus a sound supervisor and two sound engineers who rig, or set up, the **SFX** (sound effects) microphones and the commentator's microphone. This is usually a lip mic (pronounced 'mike') which picks up only the sound of his or her voice.

Round the back of the ground is the scanner. Looking like a large removal van, this is a mobile **gallery**, or control room. In it are the **director**, **vision mixer**, **production assistant** and sound supervisor. Sometimes the director does his own vision mixing. Different sports present different challenges. While

OB vehicles at Newmarket races.

Slo-mo

There are usually three slo-mo (slow motion) machines in use for replays. The director decides which cameras they will be slaved, or attached, to. This will usually be the wide angle coverage camera and the two goal cameras, but they can be switched to other cameras during a match.

Special cameras

Various special cameras are used in OB work:
• Radio camera. A camera connected to the scanner by a radio signal rather than a cable. One may be mounted on a jeep driven alongside the track when covering horse racing.
• Miniature cameras. Miniature radio cameras can be placed in jockeys' caps, motor racing drivers' helmets and cricket stumps.
• Steadi-cam. A Steadi-cam is a radio camera strapped to a cameraman's chest. This can be used for pitch-side interviews, shots in the players' tunnel, etc. Steadi-cam operators must be good at walking backwards. Some operators have been known to wear one for three-and-a-half hours.
• Aerial camera. Really wide-angled views can be obtained from a helicopter-mounted camera or, on a still day, from an airship-mounted camera.

DID YOU KNOW?•FACTS & FIGURES•INDUSTRY STATISTICS•VITAL STATISTICS•DID YOU KNOW?•FACTS & FIGURES•INDUSTRY STATISTICS•VITAL ST

Extra money for football from TV

• 1983: BBC AND ITV PAID £5.2 MILLION TO COVER TEN LIVE LEAGUE GAMES A YEAR FOR TWO YEARS.
• 1988: ITV PAID £44 MILLION FOR EXCLUSIVE COVERAGE OF 21 LIVE LEAGUE GAMES A YEAR FOR FOUR YEARS.
• 1996: BSKYB PAID £670 MILLION FOR EXCLUSIVE COVERAGE OF 60 LIVE GAMES OVER A FOUR-YEAR PERIOD.

Available to all?

For many years, BBC Television had the rights to show most major sporting events. ITV covered some sports but it was to the BBC that most fans turned for uninterrupted coverage (that is, for coverage uninterrupted by commercials) of cricket, golf, rugby, tennis and, of course, football.

With the arrival of satellite television, there was more competition for the rights to cover matches and races. When a satellite company bought exclusive rights to a sport or particular event, viewers without satellite or **cable** television complained because they could no longer see it.

The Government drew up a list of the so-called Crown Jewels, the most important sporting events which were always to be available to all viewers. Those on the list in 1997 can be seen across the top of these two pages. Although these may be the most important events, they were not always the ones that attracted the largest audiences. In that year, the Grand National horse race did attract the largest number of viewers (15 million) followed by the FA Cup Final (11 million). But the next eight largest audiences were all for football matches, seven of them being games involving Manchester United. No wonder the club decided to set up its own subscription channel.

BSkyB (or Sky) has three sports channels providing more than 14,000 hours of sport a year. Viewers have to pay extra for these channels.

Fly on the wall

Fancy having a TV crew in your house for six weeks, videoing everything that goes on? Including your private rows and sulks?

One of the most popular types of documentary is the fly-on-the-wall style of programme. This takes you behind the scenes and supposedly lets you see everything a fly on a wall would see if it were in the place where the series is made. One of the first such series did feature the daily life of an ordinary family. Other fly-on-the-wall series have been made about hotels, airports, driving instructors and a rugby club.

Some documentaries are more like investigations. They attempt to discover the truth about political, medical and other issues.

Unlike a drama or a comedy series, a documentary does not start with a script. The **director** is, after all, going to film a slice of real life. First, he or she must convince a senior **producer** or executive it is worth making the programme or series. For this, the director will have to write a treatment.

A treatment is usually a single sheet of A4 paper. On it will be typed:
• the working title of the film (its eventual title may be decided much later)
• its duration or length
• what series or slot in the schedules it is intended for
• a summary of what it will be about
• suggested elements (that is, people who might be interviewed, places and events that might be filmed, archive or historic film, etc.)
• how it will be made (film or video) and the number of days it will take
• the planned budget, or cost.

This is the piece of paper that must sell your idea. Only when a senior producer has bought your idea do you get given a programme number – which unlocks the budget that will allow you to make the documentary (and you can bet your budget will be less than what you asked for).

What is a documentary?

People in television give different answers but...
• A documentary illustrates or investigates a real-life subject.
• A documentary shows its subject at first hand. It is not a 'second hand' programme with people not directly involved just talking about it.
• It is a creative work. It is the way the director or director and producer or director and presenter see the subject. Their selection and ordering of shots, their editing and narration make it a personal view of the subject.
• It can vary from a short item in a programme (say, a view of one resort in a holiday programme) to a whole programme or series of programmes.

Don't forget to budget for...

• INTERVIEWEES' AND PRESENTERS' FEES
• COPYRIGHT PAYMENTS ON ARCHIVE MATERIALS
• TRAVEL, MEALS AND HOTELS FOR THE CREW
• FILM AND FILM PROCESSING OR VIDEOTAPE
• HIRE OF CAMERAS, LIGHTING, ETC.
• THE CREW'S FEES
• MUSIC
• EDITING COSTS

Making the documentary

The first step is research. You must find the best people to interview (it is no good finding the person who knows a lot about the subject if they are going to be hopelessly tongue-tied in front of the camera), the best locations to film and any archive material you want to include.

Next, you must produce the first of many drafts of the script. At this stage it will be not much more than a list:

- opening titles over film with music
- still pictures with **voice-over**
- clips from interviews
- specially shot film of location with **OOV** narration...

A single camera unit filming a documentary at a school.

Now comes the recce (pronounced 'recky' – from reconnaissance) of the locations you are likely to be filming in. Will it be quiet enough to record an interview? Will you need extra lighting? Are there power supplies available? Is there anywhere for the film crew to eat and stay overnight if necessary? Is there anyone else around who would be a better interviewee? Then it's time to write a more detailed script. It still won't include anything that will be actually said in the programme – but the shape or running order of the programme will be much clearer.

Next comes the shooting script – the order in which each section will be filmed or taped. Obviously it's easier and cheaper to do all the shooting needed in one location before moving the equipment to the next – even if that means filming sections out of sequence.

On the actual shoot, you will film or tape more than you need so you can choose the best bits. A generous ratio for documentaries is 8:1. This means you film eight minutes for every minute you expect to include in the final programme.

Then comes the **editing** – either in a cutting room with a film editor, or in a VT-editing suite with a VT editor. Here you assemble the first rough cut of your programme. Then you can write the commentary, complete the final edit, add the narration and any music, and hope that someone will watch.

Some of the equipment in a VT-editing suite.

Infotainment

In one November day, you could watch two cookery competitions, an hour-long programme about turning a birdcage into a flower vase, and a fun health programme about your pecs and glutes.

With the start of daytime television and the later rapid increase in the number of television channels, a need has arisen for plenty of easy – and quick-to-make – programmes to fill the schedules. More importantly, they had to be inexpensive.

A new type of programme was created to help fill this need: programmes which were a mixture of information and entertainment – sometimes called infotainment shows. They are often made in a studio rather than on location, which means they are quicker to make. Indeed, several editions of one show can be recorded in a day – all on the same set and (if there is one) with the same studio audience. (When one programme has been recorded, the audience will be told to remember 'it's now next week'.)

The first famous infotainment show, *That's Life*, began around 1972 and was popular for over 20 years. It was a mixture of items ranging from serious investigations into child abuse and businesses which cheated their customers, to comic items which included a dog that said 'sausages' and jokes about misprints in papers and rudely-shaped vegetables.

STUDIO SCHEDULE	
MONDAY 8TH – THURSDAY 11TH SEPTEMBER 1997	
1200	PRODUCTION TEAM / FERN
	CONTESTANTS ARRIVE
1300 – 1400	LUNCH
1330	CREW CALL
	AUDIENCE WARM UP
1400 – 1500	RECORD SHOW 1
1500 – 1530	RESET
1530 – 1630	RECORD SHOW 2
1630 – 1700	RESET
	NEW AUDIENCE IN
1700 – 1800	RECORD SHOW 3

*The **director** of a show like* Ready Steady Cook *must keep to a strict timetable if he or she is to record three editions of the show in the studio time available.*

Today, infotainment shows range from those about real-life accidents and rescues (sometimes using videotape that has been shot by the emergency services) to the many lifestyle programmes which cover topics such as gardening, clothes, decorating and, of course, cooking.

Some people think that, because they often try hard to be fun, infotainment shows are very superficial and do not address serious issues. They are sometimes quoted as examples of the 'dumbing down' of television. By this, it is meant that their **producers** are so afraid of putting viewers off, they make only programmes that are easy to follow.

The religious infotainment ITV programme, Sunday Morning, *was largely unscripted; that is, it consisted of ad lib conversations. Even so, as this script shows, the programme needed careful planning.*

Independent producers

THE BBC, ITV, CHANNEL 4 AND CHANNEL 5 ALL BUY PROGRAMMES FROM INDEPENDENT PRODUCERS. THERE ARE OVER A THOUSAND INDEPENDENT PRODUCTION COMPANIES AND YOU CAN TELL WHICH COMPANY MADE A PARTICULAR PROGRAMME BY LOOKING AT THE CLOSING CREDITS. THESE ARE THE TOP EIGHT — THE EIGHT PRODUCTION COMPANIES WHICH MAKE THE MOST PROGRAMMES:

1 CHRYSALIS
2 TRANSWORLD INTERNATIONAL
3 BROADCAST COMMUNICATIONS
4 MENTORN
5 PLANET 24
6 SUNSET AND VINE
7 HAT TRICK
8 PEARSON TV

MANY INFOTAINMENT PROGRAMMES ARE MADE BY INDEPENDENT PRODUCERS.

camera 2

on the left of the set

cameras move 'as directed' (camera operators hear the director's instructions through their headphones)

super caption generator (name is super-imposed on whichever picture is being shown)

Still pictures have been created of the covers of the three books being discussed, together with four illustrations from one of them. These pictures can be shown at suitable moments in the discussion.

in vision

centre left

running time: 4 minutes 30 seconds

```
ITEM 10
11.    CAM 2                          /KAY IN VIS
       KAY SET L                      Thanks, Graham.  And we'll be going
                                      back to Cambridge for Morning Worship
ITEM 11                               in a few minutes.  Richard.../
12.    CAM 3                                          DAVID CENTRE L
       MS RICHARD Centre set                          RICHARD CENTRE R

                                      RICHARD IN VIS
                                      This week, no less than sixty new
                                      religious books will be published –
                                      and that's pretty much par for the
13.    CAM 4                          course. /  David Self is a critic,
       MS DAVID SELF                  broadcaster and author of several
                                      books himself, and he's going to be
ITEM 12                               with us every fortnight to
14.    CAM 1                          recommend a few of the latest titles/
       2 / SHOT RICHARD / DAVID

       CAMS A / D                     RICHARD / DAVID
       CAM 1 – 2 / SHOT               ad lib book review
       CAM 3 – MS RICHARD
       CAM 4 – MS DAVID
                                      R /T :   4.30"
       S / CAPGEN
       -------------------------------------------------
       DAVID SELF
15.    STILL STORE                /
       1.  WHY DO PEOPLE SUFFER? Book
       2.  2 / SHOT LADIES
       3.  MAN HOLDING BABY
       4.  CROSSES AT SUNSET
       5.  SEA OF GALILEE
       6.  MY LIFE Book
       7.  WHERE DID I GO WRONG

SHOT 16 ON CAM 3 NEXT
```

The chat show

Suppose a new drama costs £1 million for each episode. Suppose a documentary costs £100,000 an hour. On this scale, a chat show can cost just £15,000 an hour. Chat is cheap.

What makes television different from radio is, of course, the fact that it has pictures. Even so, one of the most frequent types of programme is the chat show, or a programme which consists simply of talking heads. The reason is partly cost. The guests may be so glad to appear they do not want large fees. The simple studio set can remain unaltered from week to week and there are rarely ever **outside broadcasts**.

Chat shows are also comparatively easy to organize. There is, it seems, no shortage of people willing to appear in order to advertise or plug their latest book, film, album or TV programme. As one critic wrote, 'Their reason for turning up is so clearly to plug their wares that it is hard to tell where the show ends and the commercial break begins.'

Chat shows vary enormously. There are the proper discussions in which a number of guests talk about a newsworthy topic seriously and intelligently. There are the political talk shows in which a **presenter** interrogates one or more MPs. Then there are those in which the presenter invites showbiz stars onto his or her sofa. And there are those shows in which the guests are ordinary members of the public – and the star of the show is the presenter.

Each of these types of chat show may or may not have a studio audience. The studio audience may or may not be invited to join in. If so, it can be very interesting. On some occasions, though, the audience seems to consist only of people who want to shout down other members of the audience.

Some comic actors have made fun of the chat show format by inventing 'chat show hosts from hell'.

For instance, Carole Aherne appears as Mrs Merton, Barry Humphries as Dame Edna Everege and Steve Coogan as Alan Partridge.

Daytime television relies heavily on variations of the chat show format. Some shows will have regular guests such as gardening experts, doctors and astrologers. Regular guests mean the show is easier to organize.

Increasingly, chat shows will become interactive. Already many shows encourage members of the public to phone in and have their voices heard. In others, it is possible for viewers to vote by phone. The spread of video phones will make it even easier for the public to take part. Those shows which expect viewers to ring premium (that is, expensive) phone numbers not only get audience participation for free: they can make money out of the calls.

Didn't he do well?

People who want to plug a new product want to do it well. People who have been asked to appear on television to defend themselves also want to do well. For example, politicians want to put their arguments across effectively. Owners, directors and managers of companies that may be in the news want to create a good impression. The managers of hospitals and public service companies (providing water, gas or electricity) all need to perform convincingly on TV.

There are companies that provide training for chat show guests in 'beating the interviewer' and 'selling yourself'. These are two pages of the kind of advice such companies provide.

TELEVISION MAKEUP

Not every television station has a makeup artist. Find out beforehand. If there isn't going to be professional help and you want to look your best, here are a few guidelines:

* **ANTI-PERSPIRANT**
If necessary, apply anti-perspirant gel sparingly to the face.
* **CONCEALER**
Apply only if you need to hide any blemishes or spots.
* **FOUNDATION**
Apply over all of the face and blend towards your hair line and ears and to just under the jaw line. Don't forget the eye area.
* **POWDER**
Pat powder onto the face to reduce shine. Powder is the most important step so even if you use nothing else you must use powder. Bald men may need to powder their scalps. For women, eye shadow, eyeliner, mascara and lipstick may be desired. Remember that natural quiet colours come across far better on TV than bright cerise pink lips and bright green eyeshadow.
* **HAIR**
Check for loose hair as it will cause problems if it falls in your eyes during a live interview. Always comb or brush your hair (where styles allow). Run your hand gently over the head to reduce static. If you're being interviewed outdoors spray your hair into place. This is optional for an indoor interview.
* **CLOTHING**
Avoid wearing tiny check or herringbone suits and narrow pin-striped shirts which may 'flicker' or 'strobe' on screen. Women should avoid too many frills and jewellery which can sparkle distractingly.
* **LASTLY**
Check your shoulders for dandruff.

TELEVISION INTERVIEWS
BASIC TIPS

BEFORE THE INTERVIEW

Research: Double check and learn any facts you may need to know. Make sure you understand the arguments for and against your point of view. Make a list of questions you think you may be asked. Get help from colleagues to answer the questions you think are difficult.

Rehearsal: If you feel nervous about a television interview, try a rehearsal with a colleague asking the questions you think will come up. Don't avoid difficult questions. If you can think of them a journalist can.

Remember: Decide on your three key points. Be prepared to make them. If you can think of a good phrase that sums up your position, remember it and use it – TV producers are often looking for a short soundbite (15-20 seconds) to use in news bulletins.

DURING THE INTERVIEW

Remember: Listen to the questions but make sure you make your three points, even if it means avoiding the journalist's question.
Remain polite. Don't lose your temper. You'll come off worse in the viewer's eyes.

Producer guidelines

'Television is simply chewing gum for the eyes.'
'No. It affects everything we do, everything we think.'

Some people think that television has little effect on us. It does not change the way we behave. It should be free to show what it likes. Others think it is a hidden persuader and needs strict controls, even censorship. In fact, in this country, there are quite a few rules or guidelines about what can and can't be shown on the TV screen. The problem is, not everyone agrees with them – or how to put them into practice.

These are the kinds of questions that are difficult to resolve:
• A film star falls over, drunk, in the street. Should that be shown on the news?
• Suppose a relative of yours has been selling hard drugs. Has a documentary the right to put his or her face on screen?
• Is an interviewer right to get angry with someone he or she is certain is lying?
• If you were in hospital after being injured in a train crash, should TV be allowed to show your disfigured face or try to interview you?

BBC Producers' Guidelines

The BBC issues **producers** with a 300-page book of guidelines to answer questions like these. Here are some of the points they make on just two issues:

Interviews

• BBC interviewers 'should not appear to be sympathetic' to controversial positions. 'They should appear searching, sharp, sceptical, informed – but not ... discourteous or emotionally attached to one side of an argument.'
• BBC interviewers are told not to ask leading questions, such as: 'Are you in this mess because you are dishonest or just foolish?' But interviewers are also warned that some interviewees are skilled at avoiding questions.
• 'In a well-conducted interview ... viewers regard the interviewer as working on their behalf.'

Hidden cameras

The guidelines do not rule out all use of hidden or long-range cameras but make these points among others:
• 'People in a public place cannot expect the same degree of privacy as in their own homes. They can be seen by anyone, and that means they may be spotted by cameras.'
• 'On private property – and especially in their homes – people may reasonably expect not to be watched or listened to by the BBC.'

The media scrums Princess Diana often experienced are not ruled out by the guidelines.

• 'The BBC will generally use hidden cameras or microphones on private property only where ... evidence exists of crime or significant anti-social behaviour.'

• 'Prominent public figures must expect media attention when they become the subject of news **stories**.'

• Secret 'recording of identifiable people in grief or under extremes of stress (for example, in hospitals) requires special consideration'.

Independent Television Commission

The Independent Television Commission (ITC) licenses and regulates commercial television services provided in and from the UK. These include Channel 3 (ITV), Channel 4, Channel 5 and a range of **cable** and **satellite** services.

One of the ITC's most important functions is to see that the programmes and advertisements on the services it regulates follow its *Programme Code*. The main requirements of the code are:

• Programmes should not include material which offends against good taste or decency, is likely to incite crime, lead to disorder or be offensive to public feeling.

• News... has to be both accurate and impartial.

• Programmes dealing with controversial subject matter have to be both impartial and fair.

• Religious programmes must not misrepresent religious beliefs and practices.

• There must be an important public interest served by any intrusion into an individual's privacy.

• Commercial products or services must not be promoted (in programmes).

• Technical devices should not be used in programmes without viewers being aware of it. This last point is to prevent television stations flashing up messages (for example, 'Go out at once and buy...') for such a short period of time the viewer does not notice them but nevertheless absorbs the message. This is sometimes called subliminal advertising.

The watershed

Television channels and companies are held responsible for making sure that nothing that is unsuitable for children is shown before 9.00 pm. This time in the day is known as the watershed. The ITC *Programme Code* says: 'After 9.00 pm and until 5.30 am progressively less suitable (ie more adult) material may be shown and it may be that a programme will be acceptable for example at 10.30 pm that would not be suitable at 9.00 pm. But it is assumed that from 9.00 pm to 5.30 am parents may reasonably be expected to share responsibility for what their children are permitted to see. Violence is not the only reason why a programme may be unsuitable for family viewing. Other factors include bad language, profanity, crude innuendo, explicit sexual behaviour, and scenes of extreme distress.'

DID YOU KNOW?•DID YOU KNOW?•DID YO

Broadcasting Standards Commission

THE BSC WAS FORMED IN 1997 AND INVESTIGATES COMPLAINTS ABOUT THE CONTENTS OF PROGRAMMES (TASTE AND DECENCY, VIOLENCE) FROM ANY MEMBER OF THE PUBLIC. IT ALSO INVESTIGATES COMPLAINTS ABOUT FAIRNESS AND INVASION OF PRIVACY BUT ONLY FROM THOSE INVOLVED. ITS ADDRESS IS 7, THE SANCTUARY, LONDON SW1P 3JS.

Scheduling and the audience

When BBC 2 started in 1964, Monday nights were for family entertainment, Tuesdays were for educational programmes, Wednesdays for repeats, Thursdays for minority interests, and Fridays for family drama. That plan lasted less than three months.

The various channels are all in competition with each other to get the biggest possible audiences. If the commercial channels don't get big audiences, they will have trouble persuading advertisers to buy time in the commercial breaks. If the BBC does not attract enough viewers, it fears the Government will say it does not deserve to receive the licence fee – or that the licence fee will not be increased in future. (BBC 2 and Channel 4 are meant to be different: their controllers are usually happy if they get around 10% of the audience.)

It is not just by showing good or popular programmes that the channels attract viewers. The time a programme is shown can alter the size of its audience. For example:

• ITV programmes often begin on the hour or at half past the hour, so, on a Sunday evening, if BBC 1 starts a popular programme at 8.50 pm, its viewers may not switch over to ITV at 9.00 pm.

• ITV planners know what time the news will be on BBC 1. If they start a popular drama series or film at that time, they have a good chance of tempting a lot of viewers away.

• Some years ago, both BBC 1 and ITV showed current affairs programmes (programmes about topical, newsworthy matters) at the same time on Monday evenings. When BBC 1 moved its programme *Panorama* from 8.10 pm to 9.30 pm, it doubled its audience.

The placing or timetabling of programmes is called scheduling.

Viewing schedules

Schedulers divide the year up into four periods of thirteen weeks each and begin scheduling each quarter over a year in advance. Some programmes, such as the news, can be slotted into their regular times at once. Then the various programme-makers bid for the remaining slots – each arguing that their programmes will win big audiences. Network controllers buy or commission the ones they like best – and production can start on those selected.

By January, the look of the autumn quarter will begin to take shape. Much will still change. A series may not be ready in time. A star may not be available. Or a show that has been finished may not turn out to be suitable for its intended slot.

The last deadline for changes to the schedule is when the listings magazines (such as *Radio Times* and *TV Times*) are printed. Even then, the schedules may be changed. Whenever a channel has a disaster movie scheduled, such as a film about a plane crash, an alternative film is kept on standby just in case a real air crash happens.

Some people say that nowadays everyone is their own scheduler. With 80% of the population having access to a video recorder and therefore able to 'time-shift' programmes, with the opportunity to buy or rent videos, and with so many new channels, we have the opportunity to create our own viewing schedules.

Idents and promos

Because there are now so many channels to choose from (at least for viewers with **satellite** or **cable** television), television controllers are always keen to remind viewers which channel they are watching – and, hopefully, keep them loyal to their channel. Some channels show a tiny **ident** – or identification symbol – in a corner of the screen all the time. Others rely on showing their ident between programmes. Idents are often accompanied by a specially written tune which viewers come to associate with the channel.

Until 1997, BBC 1 used a revolving globe which didn't actually exist: it was created by a computer. Then it spent £0.5 million filming a real hot-air balloon in different locations as its new ident or logo.

Similarly, a lot of money is spent creating advertisements for future programmes. These adverts are known as promos (promotions) or trails (trailers).

Factual programme audiences

IN THE COURSE OF A YEAR, BBC FACTUAL PROGRAMMES ARE SEEN BY, OR REACH, THESE PERCENTAGES OF THE ADULT POPULATION:

ARTS AND MUSIC PROGRAMMES	55%
CURRENT AFFAIRS	37%
DOCUMENTARIES	91%
NATIONAL NEWS	70%
REGIONAL NEWS	47%
SPORT	52%

Channel ratings

IN A TYPICAL WEEK IN LATE 1997, THE MAIN CHANNELS ATTRACTED THESE VIEWERS:

	SHARE OF THE AUDIENCE	AVERAGE HOURS PER VIEWER
BBC 1	29.8%	7.5
BBC 2	11.5%	2.9
ITV	33.4%	8.4
CHANNEL 4	10.1%	2.5
CHANNEL 5	3.4%	0.9
OTHERS (SATELLITE & CABLE)	11.8%	3.0
TOTAL	100%	25.1

AUDIENCE FIGURES ARE COLLECTED BY BARB (BROADCASTERS AUDIENCE RESEARCH BOARD LTD), WHICH IS OWNED JOINTLY BY THE BBC AND ITV. IT COLLECTS ITS FIGURES BY MEANS OF METERS ATTACHED TO A SAMPLE OF 4500 FAMILIES' SETS AND BY ASKING VIEWERS TO KEEP DIARIES OF THEIR VIEWING.

These idents were created by the advertising agents Lambie-Nairn & Company.

Transmission

In the early days, a television set cost 20 times the average weekly wage. To receive a picture you needed a large aerial on the roof of your house.

Transmitters

Until comparatively recently, all television pictures were delivered to the home by carrier waves sent out by **transmitters**. These transmitters are often on hills or mountains so that the signal can reach the maximum number of homes.

The United Kingdom is covered by a network of more than 400 transmitters and booster stations, which carry television signals to about 99% of the population. The trouble with this system is that only a certain number of frequencies are available, which limits the number of channels. In some parts of the country, there are not even enough frequencies to make Channel 5 available. Because the transmitter is on the ground, this is sometimes called terrestrial television (from the Latin *terra*, meaning earth).

Satellite

News reports and programmes had been sent across the world from studio to studio by **satellite** since the 1960s, but in 1982 it became possible to receive channels at home by means of a small dish. Pioneers like the Sky Channel beamed their broadcasts to satellites above the Equator and down to a 'footprint' area which covered the whole of Europe and North Africa. Year by year, satellites became more powerful, and dishes became smaller, so that now, over three million homes in Britain have their own satellite dishes. As long as they have the right equipment, viewers in Britain can watch French, German, Italian and Spanish channels.

To take advantage of this new type of transmission, programmers launched more than a hundred channels during the 1990s, including children's, sports, rock music, film and many specialized channels. For example, the Parliamentary Channel provides continuous coverage of what is happening in Parliament.

Cable

Since 1984, television has also been delivered by **cable**. A cable provider company installs underground cables throughout a town or city, radiating from a central control point called the headend. It uses large dishes at the headend to receive various channels from satellites. It then distributes a selection of channels along the cables to homes that pay to be connected to the system. The cable system can also be used to carry ordinary telephone calls. However, cable was very slow to become popular. By 1989, fewer than 100,000 homes had been connected.

Digital

Digital television broadcasting is even more efficient. In digital, the sounds and pictures are compressed for transmission and decoded by a set-top box. Increasingly, these decoders will be built into new televisions. Digital television can be delivered by transmitter, satellite or cable. Ten separate digital channels can be carried in the space used by one traditional television signal. This enormously increases the number of free-to-viewer, subscription and pay-per-view channels. Digital television began in this country in 1998.

 One of the main providers of satellite and cable channels is British Sky Broadcasting. It offers twelve Sky channels and is a partner in thirteen joint ventures. Among its wholly-owned channels are Sky 1, Sky News, several sports channels, Sky Movies Screen 1 and 2, and Sky Soap. Its jointly owned channels include Nickelodeon, QVC (a home-shopping channel), the Granada channels (Granada Plus, Granada Men and Motors etc.), the History Channel and Sky Scottish.

 In 1997, the BBC went into partnership with a commercial company, Flextech, and formed UKTV. This is the brand name for a set of channels including UK Gold, UK Horizons (factual programmes) and UK Style (programmes about leisure and living – the Lifestyle channel). In effect this meant that the BBC was for the first time running channels funded by adverts. Any profits, however, went back into BBC programme-making.

The state of cable

NUMBER OF SEPARATE CABLE SYSTEMS	124
HOMES PASSED	8,962,929
TOTAL CONNECTIONS	2,676,897
TV CONNECTIONS	1,968,342
PHONE CONNECTIONS	2,498,646

Paying as you watch

There are two main ways to pay for extra television channels:
• Subscription: You pay an additional fee (usually per month) to receive a particular channel by satellite or cable. These include the premium movie and sports channels.
• Pay-per-view: The provider of the channel charges extra for a particular programme, such as a major boxing match.

One day, sometime in the future, non-digital transmission systems will be switched off.

Films on demand

New ways of delivering TV programmes are becoming possible. British Telecom's Asynchronous Digital Services Loop allows you to dial up a film via your phone. Similarly, it will become increasingly easy to watch programmes on your personal computer, via the Internet. You may have to pay the phone bill for as long as you are on line.

*Dishes **uplink** television signals to satellites for retransmission.*

Key dates

1932	First experimental television programme is transmitted by the BBC.
1936	BBC begins the first regular television service in the world (seen only in and around London).
1939	Television service closes down because of World War II.
1946	BBC Television Service resumes.
1949	BBC Television is transmitted in the Midlands (gradually extended to the rest of the country over the next eight years).
1950	First live broadcast from the continent.
1953	Live coverage of Queen Elizabeth II's coronation (the largest outside broadcast to date).
1955	ITV begins broadcasting in the London area (extended to other regions over the next six years).
1961	First live broadcast from Moscow; *Coronation Street* begins.
1962	First exchange of live television between Britain and America (by Telstar **satellite**).
1964	Start of BBC 2.
1966	First direct pictures from the moon.
1967	Regular colour programmes begin on BBC 2.
1968	Thames Television (broadcasting on weekdays) and London Weekend Television, Yorkshire Television and Harlech Television begin transmission (replacing ABC-TV, Rediffusion and Television Wales and West).
1969	Colour broadcasting extended to BBC 1 and ITV.
1973	ITV Oracle (teletext) service begins.
1974	BBC Ceefax (teletext) service begins.
1977	More colour than black-and-white licences issued for the first time (colour licences cost £21, black-and-white £9).
1979	ITV on strike for eleven weeks.
1981	Westward Television replaced by Television South West (TSW).
1982	Central Television and Television South (TVS) begin broadcasting (replacing ATV, Southern and Westward); Channel 4 and S4C (the Welsh channel) begin.
1983	Breakfast-time television begins on BBC and ITV (TV-am).
1989	Sky Television begins (with four channels); Cameras allowed into the House of Commons (after earlier experiments in the House of Lords).
1990	Sky Movies is Britain's first **encrypted** satellite channel; Sky becomes BSkyB.
1991	BBC World Television begins; BSkyB now offers five channels.
1993	ITV licences renewed (Thames, Television South, Television South West. and TV-am losing their franchises to Carlton, Meridian, West Country and GMTV).
1997	Channel 5 begins; BBC's first commercial channels begin (UKTV).
1998	Digital television begins.

Glossary

actuality   sound and pictures obtained on location, rather than in a studio

autocue (or prompt)   roll of script reflected in front of the camera lens, so presenters can read their script while looking into the camera

bird   jargon for **satellite**

brief   instructions given to a reporter, concerning the story he or she is to cover

bulletin   summary of the day's news, lasting no more than five minutes

cable TV   television delivered to the home by underground cable

chromakey   device whereby a still or moving picture can be shown behind a **presenter**

DBS   direct broadcast **satellite** which sends signals to domestic 'dish' aerials

director   the person in charge of making a programme

earpiece   listening device worn by a **presenter** of a programme. It is invisible to the audience but allows the **director** in the **gallery** to talk directly to the presenter.

editing (or video-editing)   electronic selection and re-arrangement of pictures and sound

ENG (electronic newsgathering)   newsgathering using video

encryption   scrambling of **satellite** transmission; the viewer must pay to decode it

floor manager   human link between the **gallery** and the studio floor. The floor manager uses headphones to listen to instructions from the **director** in the gallery.

gallery   control room in a TV studio which contains the mixing desk and monitors

ident   image or logo used to identify a television channel

intake editor   news editor responsible for commissioning news reports

link   introduction to a news **story** or location report, spoken by the studio **presenter**. Also a spoken announcement between programmes.

monitor   screen used for closed-circuit picture display. Monitors, unlike TV sets used in the home, do not receive broadcast pictures via an aerial.

news programme   longer and more detailed view of events than a **bulletin**

optical-fibre cable   thin glass fibres bunched together to make a **cable** capable of carrying a very large number of TV channels. Also called fibre optic cable.

outside broadcast (OB)   programme made outside a television studio

OOV   out of vision (see also **voice over**)

presenter   person who appears on the screen to introduce various items and guests. In North America he or she is called the anchor, frontman, or linkman.

producer   person in overall charge of the programme or series, responsible for its budget and (often) its content. He or she employs the **director** to actually make the programme.

production assistant   person responsible for programme timings in the **gallery**

prompt   see **autocue**

rigger driver   person who drives the **outside broadcast** vehicle (the scanner) and does many of the physical jobs involved in setting up an outside broadcast

satellite   communications satellites pick up TV signals beamed up from Earth, amplify them, and transmit them to receiving stations

SFX (or FX)   sound effects

special needs   a term used to describe someone who has mental or physical difficulties

story   news event covered by a reporter

transmitter   engineering centre with a tall mast which sends out television signals

two-shot   shot of two people

uplink   to beam a signal up to a satellite

voice-over (or v/o)   separate audio track, usually in the form of a commentary. The speaker of a voice-over is **OOV**.

vision mixer   person who sits at the mixing desk in the **gallery** and operates the switches controlling the picture sources

VTR   video tape recording

Index